Summary, Analysis & Review

of

Elizabeth Blackburn's and Elissa Epel's

The Telomere Effect

A Revolutionary Approach to Living Younger, Healthier, Longer

by

Instaread

Please Note

This is a summary with analysis.

Copyright © 2017 by Instaread. All rights reserved worldwide. No part of this publication may be reproduced or transmitted in any form without the prior written consent of the publisher.

Limit of Liability/Disclaimer of Warranty: The publisher and author make no representations or warranties with respect to the accuracy or completeness of these contents and disclaim all warranties such as warranties of fitness for a particular purpose. The author or publisher is not liable for any damages whatsoever. The fact that an individual or organization is referred to in this document as a citation or source of information does not imply that the author or publisher endorses the information that the individual or organization provided. This concise summary is unofficial and is not authorized, approved, licensed, or endorsed by the original book's author or publisher.

Table of Contents

Overview 5

Important People 7

Key Takeaways..8

Analysis 10

Key Takeaway 1 10

Key Takeaway 2 12

Key Takeaway 3 14

Key Takeaway 4 16

Key Takeaway 5 18

Key Takeaway 6 20

Key Takeaway 7 21

Key Takeaway 8 22

Key Takeaway 9 23

Key Takeaway 10 24

Authors' Style...25

Instaread on The Telomere Effect

Authors' Perspective 27

References 29

Overview

The Telomere Effect by Elizabeth Blackburn and Elissa Epel describes advances in the field of gerontology and presents practical information on how to apply scientifically based guidance to daily life. An indicator of health and longevity is the condition of a person's telomeres, which are protective DNA that exists as cap-like structures at the end of each chromosome. Telomeres have the capacity to accelerate or decelerate the aging process because of their critical role in cellular health. It's critical to understand how they function within the body and how they respond, for better or worse, to variables including diet, sleep hygiene, exercise, stress, emotions, and environmental toxin exposure.

Although some people like to believe that aging is entirely predestined by genetics, science has proven otherwise. Numerous studies have shown that a shift in lifestyle can radically alter the length of people's "healthspan," or the period in which they enjoy robust health. Lifestyle shifts can likewise delay the onset of a "diseasespan," or the period during which illness and chronic health conditions drastically lower quality of life. This is one explanation, beyond genetic factors, for why people age at different rates.

The authors began doing telomere research together in 2002. Their collaborative work has helped them pinpoint telomeres as a key to solving the puzzle of why age and longevity vary from person to person. Telomere research also marked a turning point in demonstrating the connections between mental and physical health. The length of telomeres appears to indicate an individual's overall health. Shorter telomeres are associated with higher levels of stress and higher incidence of ailments including depression, anxiety, and cardiovascular disease. Longer telomeres are associated with vibrant health, higher energy levels, and mental well being.

Fortunately, people can learn about the biological role of telomeres and adopt evidence-based practices to support the overall health of their own telomeres, which will result in heightened health on the individual and social levels.

Important People

Elizabeth Blackburn is an internationally renowned molecular biologist and president of the Salk Institute for Biological Studies. She is professor emerita in the Department of Biochemistry and Biophysics at the University of California, San Francisco.

Elissa Epel is a prominent researcher in the Department of Psychiatry at the University of California, San Francisco. Her research focuses on the mind-body connection including how stress reduction and the promotion of mental health can positively affect the length of telomeres and, in turn, improve overall health and longevity.

Key Takeaways

1. The DNA in telomeres provides essential protection against chromosomal damage as cells renew.

2. Telomeres have the capacity to grow, unlike other strands of DNA.

3. Long telomeres usually are associated with having heightened health, whereas short telomeres are associated with having many health problems.

4. People can encourage telomere growth by adopting positive habits, such as conscientiousness.

5. Chronic stress can adversely affect telomeres. Conscientiously choosing productive responses to stress preserves telomere health.

6. People prone to negative thinking tend to have shorter telomeres.

7. Moderate exercise can positively affect telomere length.

8. Getting good sleep can help strengthen telomeres.

9. Healthy eating supports optimal telomere length.

10. An individual's social environment affects telomere length, for better or worse.

Instaread

Thank you for purchasing this Instaread book

Download the Instaread mobile app to get unlimited text & audio summaries of bestselling books.

Visit Instaread.co to learn more.

Analysis

Key Takeaway 1

The DNA in telomeres provides essential protection against chromosomal damage as cells renew.

Analysis

Telomeres are protective caps of DNA that occur in pairs at the end of chromosomes. They contain DNA that is unlike other DNA located elsewhere in the body. The DNA embedded in telomeres doesn't contain a genetic blueprint. Instead, telomere DNA is protective; its job is to stop mutations and cell death during the cell renewal process. Telomere DNA protects against the proliferation of cancer cells.

Research on telomeres has started to seep into public consciousness. For example, actress and health advocate Cameron Diaz discussed the latest in telomere research and how it impacts aging in *The Longevity Book* (2016), a bestseller that she co-wrote with Sandra Bark on the science of aging. [1] Presumably, Diaz's audience is made up of women who want to learn about aging gracefully. In an article citing Diaz's reference to telomere research in *The Longevity Book*, writer Gabrielle Frank calls Diaz's realistic, science-based take on aging a "a much welcomed fresh perspective." [2] Frank's comment speaks to the potential application of telomere research in everyday life, which in this case stands to educate women on how they can take concrete steps to slow the aging process without resorting to costly cosmetics or ineffective treatments.

Key Takeaway 2

Telomeres have the capacity to grow, unlike other strands of DNA.

Analysis

An enzyme called telomerase protects the telomere and encourages healthy growth. This is one way in which telomeres are distinct from other forms of DNA. The scientific discovery of telomerase was a major breakthrough in the field of aging and longevity because it proved that people have more control over aging than was previously thought.

While there are many ways to encourage natural growth of telomeres, in recent years, researchers have developed medical procedures to help spur telomere growth. For example, Dr. Helen Blau, who leads Stanford University's Baxter Laboratory for Stem Cell Biology, was senior author of a 2015 article in the *Federation of American Societies for Experimental Biology Journal*, which outlined research on a new procedure that elongates telomeres rapidly and efficiently. In the Stanford experiment, Blau and her colleagues delivered an altered version of RNA to stem cells which, in turn, supported the production of telomerase. This resulted in the lengthening of telomeres. The scientists discovered that the cells treated with this procedure appeared younger than the untreated cells. This procedure also resulted in a rapid increase in cells, but only for a limited period, which prevented the cells from multiplying too quickly and becoming potentially dangerous. The limited cell growth encouraged by this procedure could be beneficial for drug testing and advancing further research on aging and longevity. [3]

Key Takeaway 3

Long telomeres usually are associated with having heightened health, whereas short telomeres are associated with having many health problems.

Analysis

Telomeres shorten as a natural part of the aging process. Telomere shortening can be accelerated by many factors, such as chronic stress, depression, lack of sleep, and cellular inflammation. Depending on a person's lifestyle, telomeres can lengthen even after a person has experienced chronic stress or eaten poorly. This means that people have some control over the length of their telomeres.

Scientists at the Salk Institute's Molecular and Cell Biology Laboratory have discovered that when the length of telomeres shows a "balance between elongation and trimming," the telomeres achieve stability, which correlates to overall health. [4] In December 2016, Salk Institute researchers published findings in *Nature Structural & Molecular Biology* that show that a telomere can, in fact, be too long. The researchers used human embryonic stem cells grown in their laboratory to see if shortened telomeres inhibited pluripotency, or the cell's ability to replicate into any number of different cell types, such as skin cells, cardiovascular cells, liver cells, and so on, which they did. [5] However, research has shown that the cellular weakness in excessively long telomeres could potentially lead to cancer. Scientists at the Salk Institute discovered that telomeres that were considered too long also inhibited cell pluripotency because of heightened fragility and DNA damage. [6]

This is one reason it's important for people not to buy over-the-counter supplements that claim to boost telomeres and telomerase. These supplements aren't regulated, and if a supplement causes too much growth or growth in the wrong cells, cellular mutations and dysfunction could occur.

Key Takeaway 4

People can encourage telomere growth by adopting positive habits, such as conscientiousness.

Analysis

Scientists have discovered that personality traits have a measurable impact on the length of telomeres and levels of telomerase, which can affect overall health for better or worse. To slow the process of aging and increase vitality, people should aim to work hard and stay focused, as conscientiousness can increase telomere length. Optimism and joy are also qualities that are associated with lengthened telomeres. Additionally, having a compassionate attitude toward oneself promotes optimal telomere length.

Many researchers have explored the link between positive mental habits and good mental health. For example, in 2012, Angus MacBeth, a clinical psychology professor at the University of Edinburgh, and Andrew Gumley, a professor of psychological therapy, published findings on the correlation between self-compassion and heightened protection from various psychopathologies, such as depression. They discovered that self-compassion—a term that describes an individual's embrace of his or her own suffering and self-directed kindness—was a key factor for subjects' success in achieving resilient and robust mental health. [7] It has been proven that poor mental health exhibited by conditions like depression and anxiety can impede telomere growth. Self-compassion that boosts mental well being can naturally affect telomere length and result in positive health outcomes.

Self-compassion is particularly important for survivors of childhood abuse, who are at risk for developing adverse health conditions and who tend to have shorter telomeres. Therapist Beverly Engel writes about the critical importance of self-compassion for abuse survivors. In her view, survivors tend to carry a weight of shame surrounding their experiences even though they may have been children when the abuse occurred. Whether the abuse was physical, sexual, emotional, or verbal, childhood abuse survivors should adopt an attitude of compassion toward themselves and avoid self-blame. Engel points out that childhood abuse often has a lifelong effect and can result in dangerous behaviors, such as recklessness, as well as addiction and becoming the victim of abuse as an adult. [8]

Key Takeaway 5

Chronic stress can adversely affect telomeres. Conscientiously choosing productive responses to stress preserves telomere health.

Analysis

Stress is a normal part of life. If the stress is a passing state, people can recover fairly easily on the physiological level if they address the source of the stress. For many people, stress can be a motivator for accomplishment. But when stress persists over a long period, its ongoing effects can cause serious damage to a person's telomeres by shortening their length and reducing telomerase levels.

Fortunately, there are many ways that people can manage stress productively and reverse its adverse health effects. For example, a widow might be grieving the loss of her husband while working extra hours to take care of her small children. Her grief becomes chronic, lasting for many years after his untimely death. The stress of single motherhood is amplified by her deep and persistent sorrow; she worries excessively about paying the bills. She might be able to see physical effects of this stress—perhaps her hair suddenly turns white or she is especially fatigued in the morning and unable to get out of bed because of her depression. But unbeknownst to her, her telomeres have shortened significantly.

Regardless of whether the woman links her diminished vitality to the length of her telomeres, she might eventually decide to reduce stress in her life, which will have a restorative effect on her telomeres and lead to better health and well being. She might decide to download a meditation app on her phone and commit to using it twice daily. She might reconfigure her work schedule so that she has more time to spend with her children, which alleviates some of her anxieties. She might apply for a higher-paying job or downsize to smaller living quarters, which would eliminate some financial strain. Although she might have spent years living in grief and stress, she can still restore the good health of her telomeres, which will result in her feeling more energetic and healthy regardless of any prior damage to them.

Key Takeaway 6

People prone to negative thinking tend to have shorter telomeres.

Analysis

Studies have shown that people who are prone to pessimism, hostility, and cynicism are likely to have shorter telomeres and compromised health. So it's in a person's best interest to adopt a practice of healthier thinking on a regular basis.

Fortunately, there is an established field of psychology called cognitive behavioral therapy which is dedicated to helping people learn more productive thinking patterns. In *Feeling Good* (1980), a popular early text on cognitive behavioral therapy, psychologist David Burns lays out a clinical approach to helping people transcend unhelpful, destructive thoughts. Burns's core belief is that thoughts precede feelings. Thus, a negative thought about a situation or person creates a negative feeling. For many, this cycle becomes self-perpetuating to the point where they can't see how negative judgments lead to negative or even toxic emotions. But, according to the theory Burns sets forth in *Feeling Good*, people can interrupt this cycle simply by becoming aware of their thoughts. In time, if they are able to choose more generous or positive thoughts and assessments, they will see that their mental well being improves. [9] This improvement will boost physical health because of its restorative effect on telomere health.

Key Takeaway 7

Moderate exercise can positively affect telomere length.

Analysis

Exercise reduces oxidative stress and combats free radicals, which are particles that reduce cells' molecules and make them less efficient. In particular, moderate aerobic exercise and high-intensity interval training, which involves short periods of full physical exertion, boost telomerase. It is not necessary to engage in extreme exercise to see health benefits.

A number of scientific studies have measured telomere length in relation to how much physical exercise a person gets. A January 2017 study published in the *American Journal of Epidemiology* focused on the exercise habits of older women to determine if sedentariness had an ill effect on their telomere length. Researchers at the University of California, San Diego School of Medicine followed approximately 1,500 women between the ages of 64 and 95. The researchers measured how much exercise each woman completed. They found that women who got fewer than 30 minutes of daily exercise displayed damaged and shortened telomeres. If a woman sat for a long stretch of time, but managed to get a minimum of 30 minutes of exercise per day, she displayed significantly better cellular health. [10]

Key Takeaway 8

Getting good sleep can help strengthen telomeres.

Analysis

People who generally sleep well have stable telomere length. To ensure better quality sleep, people should get into a somewhat predictable sleep rhythm by going to bed and waking up around the same times every day. Length of sleep is key; most people need a minimum of seven hours daily.

Good sleep hygiene is within everyone's reach. For example, a person with insomnia can encourage sleep with natural methods, such as taking a melatonin supplement or consuming melatonin in foods where it occurs naturally, such as cherries. In 2012, researchers published findings from an experiment in which they gave healthy men and women between the ages of 18 and 40 a glass of tart cherry juice in the morning and at night. They found the consumption of cherry juice correlated to longer sleep time and better sleep quality. [11] Cherry juice is thus indirectly linked to better telomere health.

Key Takeaway 9

Healthy eating supports optimal telomere length.

Analysis

A healthy diet is an essential part of slowing the aging process. Consuming a balanced, whole-foods diet successfully guards against chronic inflammation, which occurs when the immune system is responding to a threat; insulin resistance, which causes people to have difficulty regulating glucose levels in the body; and oxidative stress, which occurs when free radicals drastically reduce cellular molecules. Each of these conditions can impede telomere restoration and growth.

A middle-aged man who wants to slow the aging process and feel more healthy and robust might be tempted by a trendy juice cleanse or detox. Perhaps he'll follow the latest celebrity fad diet under the delusion that eating like celebrities eat will guarantee glowing skin. However, his best decision would be to skip the high-concept diets and opt for a more simple but solid food plan, such as a diet that includes leafy greens, fruits, whole grains, legumes, nuts, and Omega-3 fatty acids like those found in nuts or fish.

Key Takeaway 10

An individual's social environment affects telomere length, for better or worse.

Analysis

Every aspect of people's health can support healthy cellular renewal including how they feel about their living arrangement, their neighborhood, and their relationships. Therefore, people should do their best to create more welcoming communities, take safety measures to prevent constant worries, and strengthen their social ties.

Without directly citing telomere length, researchers have already established that stress from unhealthy relationships can adversely affect physical health. For example, in 2012, two researchers from UCLA, Steve Cole and Naomi Eisenberger, published a study in *Nature Neuroscience* that confirmed a link between social ties and physical health as measured by neurocognitive processes. They found that when a social link is somehow threatened, it prompts a neural reaction similar to that which would occur if the person felt physically threatened. [12] These stress responses can adversely affect a person's telomeres and worsen overall health.

Authors' Style

Elizabeth Blackburn and Elissa Epel write engagingly and authoritatively about scientific advancements in the field of aging and longevity with a focus on telomeres and the protective function that telomeres perform. Burns and Epel begin the book with a hypothetical example that raises the question of whether aging and incidence of disease is entirely genetic or people have some degree of control over the process. The authors are enthusiastic about delivering the good news: people do have some control. This sets up the rest of the book as a way to help readers understand the physiological processes that cause natural aging and the specific, practical ways in which they can boost their own health and slow the aging process.

The authors are scientific researchers but write about complex biological functions in accessible language. They refer to numerous studies and emphasize that the information they share is based in science. They acknowledge areas where the science remains uncertain.

The book is set up into four parts. The first outlines what telomeres are; the second explains the effect that the mind has on a person's cells. The third part suggests practical ways to boost telomere growth, while the fourth part indicates how a person's social environment can affect telomeres. The authors use an illustration of shoelaces throughout the book and direct readers to use that visual cue as a reminder to pause and be mindful whenever they come across it. The book includes exercises that the authors call Renewal Labs. These exercises offer ways that people can apply telomere research to their everyday lives. Throughout the book, illustrations are used to help explain content. A Telomere Manifesto breaks down the essential knowledge about telomeres into easy-to-digest bullet points. The authors include their thoughts on the available tests to measure telomere length and indicate how much such tests might cost.

Authors' Perspective

Both authors are leading researchers in gerontology. Elizabeth Blackburn's research on telomeres began in the early 1980s when she sequenced the ends of DNA in small pond-dwelling organisms. Her work is the foundation of the entire field of telomere research and is responsible for the discovery that cellular aging is not strictly determined by genetics. Along with two colleagues, she won the Nobel Prize in 2009 for her work on telomeres and telomerase. Elissa Epel is a highly accomplished health psychologist who specializes in aging, stress, obesity, and the mind-body connection.

The authors believe that having individuals become more educated about telomeres can ultimately improve the state of the world because all beings are connected. Improving individual health, they believe, helps promote collective wellness.

The authors are co-owners of Telomere Diagnostics, a for-profit company that performs molecular testing.

~~~~ END OF INSTAREAD ~~~~

Instaread

Thank you for purchasing this Instaread book

Download the Instaread mobile app to get unlimited text & audio summaries of bestselling books.

Visit Instaread.co
to learn more.

References

1. "Bestselling Books, Health." *New York Times.* May 2016. Accessed January 24, 2017. http://www.nytimes.com/books/best-sellers/2016/05/15/health/

2. Frank, Gabrielle. "Cameron Diaz Is Ready To Start A Conversation On Aging." *Mind, Body, Green.* January 16, 2016. Accessed January 19, 2017. http://www.mindbodygreen.com/0-23393/cameron-diaz-is-ready-to-start-a-conversation-on-aging.html

3. Ramunas, J., et al. "Transient delivery of modified mRNA encoding TERT rapidly extends telomeres in human cells." *Federation of American Societies for Experimental Biology Journal.* 29.5 (May 2015): 1930-9. Accessed January 17, 2017. https://www.ncbi.nlm.nih.gov/pubmed/25614443

4. Rivera, Teresa, et al. "A balance between elongation and trimming regulates telomere stability in stem cells." *Nature Structural & Molecular Biology.* 24 (January 2017): 30-39. Accessed January 17, 2017. http://www.nature.com/nsmb/journal/v24/n1/full/nsmb.3335.html

5. Ibid.

6. Ibid.

7. MacBeth, Angus, and Andrew Gumley. "Exploring compassion: A meta-analysis of the association between self-compassion and psychopathology." *Clinical Psychology Review.* 32.6 (August 2012): 545-552. Accessed January 19, 2017. http://www.sciencedirect.com/science/article/pii/S027273581200092X

8. Engel, Beverly. "Healing the Shame of Childhood Abuse Through Self-Compassion." *Psychology Today.* January 15, 2015. Accessed January 19, 2017. https://www.psychologytoday.com/blog/the-compassion-chronicles/201501/healing-the-shame-childhood-abuse-through-self-compassion

9. Burns, David. *Feeling Good: The New Mood Therapy.* New York: Avon, 1980, 1999. Amazon Kindle edition, loc. 657.

10. Shadyab, Aladdin, et al. "Associations of Accelerometer-Measured and Self-Reported Sedentary Time With Leukocyte Telomere Length in Older Women." *American Journal of Epidemiology* (January 18, 2017): 172-184. Accessed January 19, 2017. https://academic.oup.com/aje/article/doi/10.1093/aje/kww196/2915786/Associations-of-Accelerometer-Measured-and-Self

11. Howatson, G., et al. "Effect of tart cherry juice (Prunus cerasus) on melatonin levels and enhanced sleep quality." *The European Journal*

of Nutrition 51.8 (December 2012): 909-16. Accessed January 19, 2017. http://www.ncbi.nlm.nih.gov/pubmed/22038497

12. Eisenberger, Naomi, and Steve Cole. "Social neuroscience and health: neurophysiological mechanisms linking social ties with physical health." *Nature Neuroscience.* 15.5 (May 2012): 669-674. http://www.nature.com/neuro/journal/v15/n5/full/nn.3086.html

Lightning Source UK Ltd.
Milton Keynes UK
UKHW01f0858221018
330966UK00009B/296/P